GEOMETRICAL SOLUTIONS

DERIVED FROM
MECHANICS

A TREATISE OF ARCHIMEDES

RECENTLY DISCOVERED AND
TRANSLATED FROM THE GREEK BY
DR. J. L. HEIBERG
PROFESSOR OF CLASSICAL PHILOLOGY

AT THE UNIVERSITY OF COPENHAGEN

WITH AN INTRODUCTION BY
DAVID EUGENE SMITH
PRESIDENT OF TEACHERS COLLEGE,
COLUMBIA UNIVERSITY, NEW YORK

ENGLISH VERSION TRANSLATED FROM THE GERMAN
BY LYDIA G. ROBINSON
AND REPRINTED FROM "THE MONIST," APRIL, 1909

CHICAGO
THE OPEN COURT PUBLISHING COMPANY
LONDON AGENTS
KEGAN PAUL, TRENCH, TRÜBNER & CO., LTD.
1909

COPYRIGHT BY
THE OPEN COURT PUBLISHING CO.
1909

INTRODUCTION.

IF there ever was a case of appropriateness in discovery, the finding of this manuscript in the summer of 1906 was one. In the first place it was appropriate that the discovery should be made in Constantinople, since it was here that the West received its first manuscripts of the other extant works, nine in number, of the great Syracusan. It was furthermore appropriate that the discovery should be made by Professor Heiberg, *facilis princeps* among all workers in the field of editing the classics of Greek mathematics, and an indefatigable searcher of the libraries of Europe for manuscripts to aid him in perfecting his labors. And finally it was most appropriate that this work should appear at a time when the affiliation of pure and applied mathematics is becoming so generally recognized all over the world. We are sometimes led to feel, in considering isolated cases, that the great contributors of the past have worked in the field of pure mathematics alone, and the saying of Plutarch that Archimedes felt that "every kind of art connected with daily needs was ignoble and vulgar"[1] may have strengthened this feeling. It therefore assists us in properly orientating ourselves to read another treatise from the greatest mathematician of antiquity that sets clearly before us his indebtedness to the mechanical applications of his subject.

Not the least interesting of the passages in the manuscript

---

[1] Marcellus, 17.

is the first line, the greeting to Eratosthenes. It is well known, on the testimony of Diodoros his countryman, that Archimedes studied in Alexandria, and the latter frequently makes mention of Konon of Samos whom he knew there, probably as a teacher, and to whom he was indebted for the suggestion of the spiral that bears his name. It is also related, this time by Proclos, that Eratosthenes was a contemporary of Archimedes, and if the testimony of so late a writer as Tzetzes, who lived in the twelfth century, may be taken as valid, the former was eleven years the junior of the great Sicilian. Until now, however, we have had nothing definite to show that the two were ever acquainted. The great Alexandrian savant,—poet, geographer, arithmetician,—affectionately called by the students Pentathlos, the champion in five sports,[2] selected by Ptolemy Euergetes to succeed his master, Kallimachos the poet, as head of the great Library,—this man, the most renowned of his time in Alexandria, could hardly have been a teacher of Archimedes, nor yet the fellow student of one who was so much his senior. It is more probable that they were friends in the later days when Archimedes was received as a savant rather than as a learner, and this is borne out by the statement at the close of proposition I which refers to one of his earlier works, showing that this particular treatise was a late one. This reference being to one of the two works dedicated

---

[2]His nickname of *Beta* is well known, possibly because his lecture room was number 2.

to Dositheos of Kolonos,[3] and one of these (*De lineis spiralibus*) referring to an earlier treatise sent to Konon,[4] we are led to believe that this was one of the latest works of Archimedes and that Eratosthenes was a friend of his mature years, although one of long standing. The statement that the preliminary propositions were sent "some time ago" bears out this idea of a considerable duration of friendship, and the idea that more or less correspondence had resulted from this communication may be inferred by the statement that he saw, as he had previously said, that Eratosthenes was "a capable scholar and a prominent teacher of philosophy," and also that he understood "how to value a mathematical method of investigation when the opportunity offered." We have, then, new light upon the relations between these two men, the leaders among the learned of their day.

A second feature of much interest in the treatise is the intimate view that we have into the workings of the mind of the author. It must always be remembered that Archimedes was primarily a discoverer, and not primarily a compiler as were Euclid, Apollonios, and Nicomachos. Therefore to have him follow up his first communication of theorems to Eratosthenes by a statement of his mental processes in reaching his conclusions is not merely a contribution to mathematics but one to education as well. Particularly is this true in

---

[3] We know little of his works, none of which are extant. Geminos and Ptolemy refer to certain observations made by him in 200 B. C., twelve years after the death of Archimedes. Pliny also mentions him.

[4] Τῶν ποτὶ Κόνωνα ἀπυσταλέντων θεωρημάτων.

the following statement, which may well be kept in mind in the present day: "I have thought it well to analyse and lay down for you in this same book a peculiar method by means of which it will be possible for you to derive instruction as to how certain mathematical questions may be investigated by means of mechanics. And I am convinced that this is equally profitable in demonstrating a proposition itself; for much that was made evident to me through the medium of mechanics was later proved by means of geometry, because the treatment by the former method had not yet been established by way of a demonstration. For of course it is easier to establish a proof if one has in this way previously obtained a conception of the questions, than for him to seek it without such a preliminary notion.... Indeed I assume that some one among the investigators of to-day or in the future will discover by the method here set forth still other propositions which have not yet occurred to us." Perhaps in all the history of mathematics no such prophetic truth was ever put into words. It would almost seem as if Archimedes must have seen as in a vision the methods of Galileo, Cavalieri, Pascal, Newton, and many of the other great makers of the mathematics of the Renaissance and the present time.

The first proposition concerns the quadrature of the parabola, a subject treated at length in one of his earlier communications to Dositheos.[5] He gives a digest of the treatment, but with the warning that the proof is not com-

---

[5] Τετραγωνισμὸς παραβολῆς.

plete, as it is in his special work upon the subject. He has, in fact, summarized propositions VII–XVII of his communication to Dositheos, omitting the geometric treatment of propositions XVIII–XXIV. One thing that he does not state, here or in any of his works, is where the idea of center of gravity[6] started. It was certainly a common notion in his day, for he often uses it without defining it. It appears in Euclid's[7] time, but how much earlier we cannot as yet say.

Proposition II states no new fact. Essentially it means that if a sphere, cylinder, and cone (always circular) have the same radius, r, and the altitude of the cone is r and that of the cylinder $2r$, then the volumes will be as $4 : 1 : 6$, which is true, since they are respectively $\frac{4}{3}\pi r^3$, $\frac{1}{3}\pi r^3$, and $2\pi r^3$. The interesting thing, however, is the method pursued, the derivation of geometric truths from principles of mechanics. There is, too, in every sentence, a little suggestion of Cavalieri, an anticipation by nearly two thousand years of the work of the greatest immediate precursor of Newton. And the geometric imagination that Archimedes shows in the last sentence is also noteworthy as one of the interesting features of this work: "After I had thus perceived that a sphere is four times as large as the cone... it occurred to me that the surface of a sphere is four times as great as its largest circle, in which I proceeded from the idea that just as a circle is

---

[6] Κέντρα βαρῶν, for "barycentric" is a very old term.

[7] At any rate in the anonymous fragment *De levi et ponderoso*, sometimes attributed to him.

equal to a triangle whose base is the periphery of the circle, and whose altitude is equal to its radius, so a sphere is equal to a cone whose base is the same as the surface of the sphere and whose altitude is equal to the radius of the sphere." As a bit of generalization this throws a good deal of light on the workings of Archimedes's mind.

In proposition III he considers the volume of a spheroid, which he had already treated more fully in one of his letters to Dositheos,[8] and which contains nothing new from a mathematical standpoint. Indeed it is the method rather than the conclusion that is interesting in such of the subsequent propositions as relate to mensuration. Proposition V deals with the center of gravity of a segment of a conoid, and proposition VI with the center of gravity of a hemisphere, thus carrying into solid geometry the work of Archimedes on the equilibrium of planes and on their centers of gravity.[9] The general method is that already known in the treatise mentioned, and this is followed through proposition X.

Proposition XI is the interesting case of a segment of a right cylinder cut off by a plane through the center of the lower base and tangent to the upper one. He shows this to equal one-sixth of the square prism that circumscribes the cylinder. This is well known to us through the formula $v = 2r^2h/3$, the volume of the prism being $4r^2h$, and requires a knowledge of the center of gravity of the cylindric

---

[8] Περὶ κωνοειδέων καὶ σφαιροειδέων.
[9] Ἐπιπέδων ἰσορροπιῶν ἢ κέντρα βαρῶν ἐπιπέδων.

section in question. Archimedes is, so far as we know, the first to state this result, and he obtains it by his usual method of the skilful balancing of sections. There are several lacunae in the demonstration, but enough of it remains to show the ingenuity of the general plan. The culminating interest from the mathematical standpoint lies in proposition XIII, where Archimedes reduces the whole question to that of the quadrature of the parabola. He shows that a fourth of the circumscribed prism is to the segment of the cylinder as the semi-base of the prism is to the parabola inscribed in the semi-base; that is, that $\frac{1}{4}p : v = \frac{1}{2}b : (\frac{2}{3} \cdot \frac{1}{2}b)$, whence $v = \frac{1}{6}p$. Proposition XIV is incomplete, but it is the conclusion of the two preceding propositions.

In general, therefore, the greatest value of the work lies in the following:

1. It throws light upon the hitherto only suspected relations of Archimedes and Eratosthenes.

2. It shows the working of the mind of Archimedes in the discovery of mathematical truths, showing that he often obtained his results by intuition or even by measurement, rather than by an analytic form of reasoning, verifying these results later by strict analysis.

3. It expresses definitely the fact that Archimedes was the discoverer of those properties relating to the sphere and cylinder that have been attributed to him and that are given in his other works without a definite statement of their authorship.

4. It shows that Archimedes was the first to state the

volume of the cylinder segment mentioned, and it gives an interesting description of the mechanical method by which he arrived at his result.

DAVID EUGENE SMITH.
TEACHERS COLLEGE, COLUMBIA UNIVERSITY.

GEOMETRICAL SOLUTIONS DERIVED FROM MECHANICS.

Archimedes to Eratosthenes, Greeting:

Some time ago I sent you some theorems I had discovered, writing down only the propositions because I wished you to find their demonstrations which had not been given. The propositions of the theorems which I sent you were the following:

1. If in a perpendicular prism with a parallelogram[1] for base a cylinder is inscribed which has its bases in the opposite parallelograms[1] and its surface touching the other planes of the prism, and if a plane is passed through the center of the circle that is the base of the cylinder and one side of the square lying in the opposite plane, then that plane will cut off from the cylinder a section which is bounded by two planes, the intersecting plane and the one in which the base of the cylinder lies, and also by as much of the surface of the cylinder as lies between these same planes; and the detached section of the cylinder is $\frac{1}{6}$ of the whole prism.

2. If in a cube a cylinder is inscribed whose bases lie in opposite parallelograms[1] and whose surface touches the other four planes, and if in the same cube a second cylinder is inscribed whose bases lie in two other parallelograms[1] and whose surface touches the four other planes, then the body enclosed by the surface of the cylinder and comprehended

---
[1] This must mean a square.

within both cylinders will be equal to $\frac{2}{3}$ of the whole cube.

These propositions differ essentially from those formerly discovered; for then we compared those bodies (conoids, spheroids and their segments) with the volume of cones and cylinders but none of them was found to be equal to a body enclosed by planes. Each of these bodies, on the other hand, which are enclosed by two planes and cylindrical surfaces is found to be equal to a body enclosed by planes. The demonstration of these propositions I am accordingly sending to you in this book.

Since I see, however, as I have previously said, that you are a capable scholar and a prominent teacher of philosophy, and also that you understand how to value a mathematical method of investigation when the opportunity is offered, I have thought it well to analyze and lay down for you in this same book a peculiar method by means of which it will be possible for you to derive instruction as to how certain mathematical questions may be investigated by means of mechanics. And I am convinced that this is equally profitable in demonstrating a proposition itself; for much that was made evident to me through the medium of mechanics was later proved by means of geometry because the treatment by the former method had not yet been established by way of a demonstration. For of course it is easier to establish a proof if one has in this way previously obtained a conception of the questions, than for him to seek it without such a preliminary notion. Thus in the familiar propositions the demonstrations of which Eudoxos was the first to discover, namely that a cone and a pyramid are one third the size of that cylinder and prism respectively that have the same base and altitude, no little credit is due to Democritos who was the first to make that statement about these bodies without any demonstration. But we are in a position to have found the present proposition in the same way as the earlier one; and I have decided to write down and make known the method partly because we have already talked about it heretofore and so no one would think that we were spreading abroad idle talk, and partly in the conviction that by this means we are obtaining no slight advantage for mathematics, for indeed I assume that some one among the investigators of to-day or in the future will discover by the method here set forth still other propositions which have not yet occurred to us.

In the first place we will now explain what was also first made clear to us through mechanics, namely that a segment of a parabola is $\frac{4}{3}$ of the triangle possessing the same base and equal altitude; following which we will explain in order the particular propositions discovered by the above mentioned method; and in the last part of the book we will present the geometrical demonstrations of the propositions.[2]

1. If one magnitude is taken away from another magnitude and the same point is the center of gravity both of the

---

[2]In his "Commentar," Professor Zeuthen calls attention to the fact that it was already known from Heron's recently discovered *Metrica* that these propositions were contained in this treatise, and Professor Heiberg made the same comment in *Hermes*.—Tr.

GEOMETRICAL SOLUTIONS DERIVED FROM MECHANICS. 12

whole and of the part removed, then the same point is the center of gravity of the remaining portion.

2. If one magnitude is taken away from another magnitude and the center of gravity of the whole and of the part removed is not the same point, the center of gravity of the remaining portion may be found by prolonging the straight line which connects the centers of gravity of the whole and of the part removed, and setting off upon it another straight line which bears the same ratio to the straight line between the aforesaid centers of gravity, as the weight of the magnitude which has been taken away bears to the weight of the one remaining [*De plan. aequil.* I, 8].

3. If the centers of gravity of any number of magnitudes lie upon the same straight line, then will the center of gravity of all the magnitudes combined lie also upon the same straight line [Cf. *ibid.* I, 5].

4. The center of gravity of a straight line is the center of that line [Cf. *ibid.* I, 4].

5. The center of gravity of a triangle is the point in which the straight lines drawn from the angles of a triangle to the centers of the opposite sides intersect [*Ibid.* I, 14].

6. The center of gravity of a parallelogram is the point where its diagonals meet [*Ibid.* I, 10].

7. The center of gravity [of a circle] is the center [of that circle].

8. The center of gravity of a cylinder [is the center of its axis].

9. The center of gravity of a prism is the center of its

axis.

10. The center of gravity of a cone so divides its axis that the section at the vertex is three times as great as the remainder.

11. Moreover together with the exercise here laid down I will make use of the following proposition:

If any number of magnitudes stand in the same ratio to the same number of other magnitudes which correspond pair by pair, and if either all or some of the former magnitudes stand in any ratio whatever to other magnitudes, and the latter in the same ratio to the corresponding ones, then the sum of the magnitudes of the first series will bear the same ratio to the sum of those taken from the third series as the sum of those of the second series bears to the sum of those taken from the fourth series [*De Conoid.* I].

I.

Let $\alpha\beta\gamma$ [Fig. 1] be the segment of a parabola bounded by the straight line $\alpha\gamma$ and the parabola $\alpha\beta\gamma$. Let $\alpha\gamma$ be bisected at δ, $\delta\beta\epsilon$ being parallel to the diameter, and draw $\alpha\beta$, and $\beta\gamma$. Then the segment $\alpha\beta\gamma$ will be $\frac{4}{3}$ as great as the triangle $\alpha\beta\gamma$.

From the points α and γ draw $\alpha\zeta \parallel \delta\beta\epsilon$, and the tangent $\gamma\zeta$; produce [$\gamma\beta$ to κ, and make $\kappa\theta = \gamma\kappa$]. Think of $\gamma\theta$ as a scale-beam with the center at κ and let $\mu\xi$ be any straight line whatever $\parallel \epsilon\delta$. Now since $\gamma\beta\alpha$ is a parabola, $\gamma\zeta$ a tangent and $\gamma\delta$ an ordinate, then $\epsilon\beta = \beta\delta$; for this

GEOMETRICAL SOLUTIONS DERIVED FROM MECHANICS. 14

indeed has been proved in the Elements [i. e., of conic sections, cf. *Quadr. parab.* 2]. For this reason and because $\zeta\alpha$ and $\mu\xi \parallel \epsilon\delta$, $\mu\nu = \nu\xi$, and $\zeta\kappa = \kappa\alpha$. And because $\gamma\alpha : \alpha\xi = \mu\xi : \xi o$ (for this is shown in a corollary, [cf. *Quadr. parab.* 5]), $\gamma\alpha : \alpha\xi = \gamma\kappa : \kappa\nu$; and $\gamma\kappa = \kappa\theta$, therefore $\theta\kappa : \kappa\nu = \mu\xi : \xi o$. And because ν is the center of gravity of the straight line $\mu\xi$, since $\mu\nu = \nu\xi$, then if we make $\tau\eta = \xi o$ and θ as its center of gravity so that $\tau\theta = \theta\eta$, the straight line $\tau\theta\eta$ will be in equilibrium with $\mu\xi$ in its present position because $\theta\nu$ is divided in inverse proportion to the weights $\tau\eta$ and $\mu\xi$, and $\theta\kappa : \kappa\nu = \mu\xi : \eta\tau$; therefore κ is the center of gravity of the combined weight of the two. In the

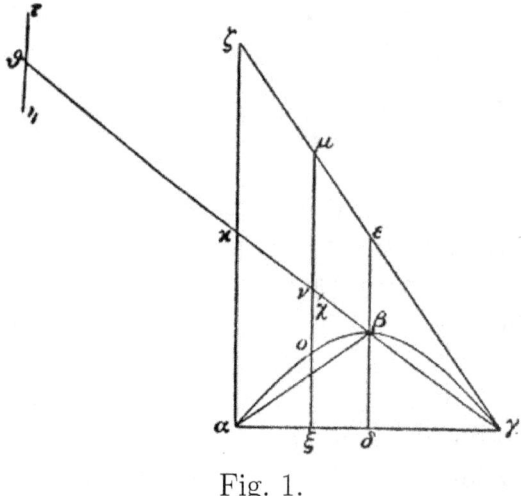

Fig. 1.

same way all straight lines drawn in the triangle $\zeta\alpha\gamma \parallel \epsilon\delta$

are in their present positions in equilibrium with their parts cut off by the parabola, when these are transferred to θ, so that κ is the center of gravity of the combined weight of the two. And because the triangle $\gamma\zeta\alpha$ consists of the straight lines in the triangle $\gamma\zeta\alpha$ and the segment $\alpha\beta\gamma$ consists of those straight lines within the segment of the parabola corresponding to the straight line ξo, therefore the triangle $\zeta\alpha\gamma$ in its present position will be in equilibrium at the point κ with the parabola-segment when this is transferred to θ as its center of gravity, so that κ is the center of gravity of the combined weights of the two. Now let $\gamma\kappa$ be so divided at χ that $\gamma\kappa = 3\kappa\chi$; then χ will be the center of gravity of the triangle $\alpha\zeta\gamma$, for this has been shown in the Statics [cf. *De plan. aequil.* I, 15, p. 186, 3 with Eutokios, S. 320, 5ff.]. Now the triangle $\zeta\alpha\gamma$ in its present position is in equilibrium at the point κ with the segment $\beta\alpha\gamma$ when this is transferred to θ as its center of gravity, and the center of gravity of the triangle $\zeta\alpha\gamma$ is χ; hence triangle $\alpha\zeta\gamma$: segm. $\alpha\beta\gamma$ when transferred to θ as its center of gravity $= \theta\kappa : \kappa\chi$. But $\theta\kappa = 3\kappa\chi$; hence also triangle $\alpha\zeta\gamma = 3\,\text{segm. } \alpha\beta\gamma$. But it is also true that triangle $\zeta\alpha\gamma = 4\triangle\alpha\beta\gamma$ because $\zeta\kappa = \kappa\alpha$ and $\alpha\delta = \delta\gamma$; hence segm. $\alpha\beta\gamma = \frac{4}{3}$ the triangle $\alpha\beta\gamma$. This is of course clear.

It is true that this is not proved by what we have said here; but it indicates that the result is correct. And so, as we have just seen that it has not been proved but rather conjectured that the result is correct we have devised a geometrical demonstration which we made known some time

GEOMETRICAL SOLUTIONS DERIVED FROM MECHANICS. 16

ago and will again bring forward farther on.

II.

That a sphere is four times as large as a cone whose base is equal to the largest circle of the sphere and whose altitude is equal to the radius of the sphere, and that a cylinder whose base is equal to the largest circle of the sphere and whose altitude is equal to the diameter of the circle is one and a half times as large as the sphere, may be seen by the present method in the following way:

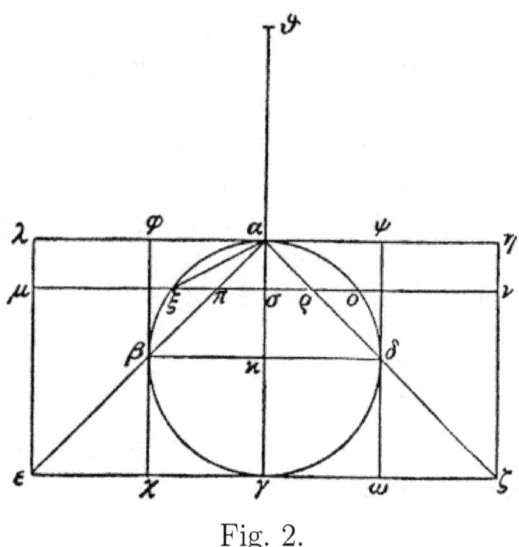

Fig. 2.

Let $\alpha\beta\gamma\delta$ [Fig. 2] be the largest circle of a sphere and

GEOMETRICAL SOLUTIONS DERIVED FROM MECHANICS. 17

$\alpha\gamma$ and $\beta\delta$ its diameters perpendicular to each other; let there be in the sphere a circle on the diameter $\beta\delta$ perpendicular to the circle $\alpha\beta\gamma\delta$, and on this perpendicular circle let there be a cone erected with its vertex at α; producing the convex surface of the cone, let it be cut through γ by a plane parallel to its base; the result will be the circle perpendicular to $\alpha\gamma$ whose diameter will be $\epsilon\zeta$. On this circle erect a cylinder whose axis = $\alpha\gamma$ and whose vertical boundaries are $\epsilon\lambda$ and $\zeta\eta$. Produce $\gamma\alpha$ making $\alpha\theta = \gamma\alpha$ and think of $\gamma\theta$ as a scale-beam with its center at α. Then let $\mu\nu$ be any straight line whatever drawn $\parallel \beta\delta$ intersecting the circle $\alpha\beta\gamma\delta$ in ξ and o, the diameter $\alpha\gamma$ in σ, the straight line $\alpha\epsilon$ in π and $\alpha\zeta$ in ρ, and on the straight line $\mu\nu$ construct a plane perpendicular to $\alpha\gamma$; it will intersect the cylinder in a circle on the diameter $\mu\nu$; the sphere $\alpha\beta\gamma\delta$, in a circle on the diameter ξo; the cone $\alpha\epsilon\zeta$ in a circle on the diameter $\pi\rho$. Now because $\gamma\alpha \times \alpha\sigma = \mu\sigma \times \sigma\pi$ (for $\alpha\gamma = \sigma\mu$, $\alpha\sigma = \pi\sigma$), and $\gamma\alpha \times \alpha\sigma = \alpha\xi^2 = \xi\sigma^2 + \alpha\pi^2$ then $\mu\sigma \times \sigma\pi = \xi\sigma^2 + \sigma\pi^2$. Moreover, because $\gamma\alpha : \alpha\sigma = \mu\sigma : \sigma\pi$ and $\gamma\alpha = \alpha\theta$, therefore $\theta\alpha : \alpha\sigma = \mu\sigma : \sigma\pi = \mu\sigma^2 : \mu\sigma \times \sigma\pi$. But it has been proved that $\xi\sigma^2 + \sigma\pi^2 = \mu\sigma \times \sigma\pi$; hence $\alpha\theta : \alpha\sigma = \mu\sigma^2 : \xi\sigma^2 + \sigma\pi^2$. But it is true that $\mu\sigma^2 : \xi\sigma^2 + \sigma\pi^2 = \mu\nu^2 : \xi\alpha^2 + \pi\rho^2 = $ the circle in the cylinder whose diameter is $\mu\nu$: the circle in the cone whose diameter is $\pi\rho$ + the circle in the sphere whose diameter is ξo; hence $\theta\alpha : \alpha\sigma = $ the circle in the cylinder : the circle in the sphere + the circle in the cone. Therefore the circle in the cylinder in its present position will be in equilibrium at the point α with the two circles whose diameters are ξo and $\pi \rho$, if they are so transferred to θ that θ is the center of gravity of both. In the same way it can be shown that when another straight line is drawn in the parallelogram $\xi \lambda \parallel \epsilon \zeta$, and upon it a plane is erected perpendicular to $\alpha \gamma$, the circle produced in the cylinder in its present position will be in equilibrium at the point α with the two circles produced in the sphere and the cone when they are transferred and so arranged on the scale-beam at the point θ that θ is the center of gravity of both. Therefore if cylinder, sphere and cone are filled up with such circles then the cylinder in its present position will be in equilibrium at the point α with the sphere and the cone together, if they are transferred and so arranged on the scale-beam at the point θ that θ is the center of gravity of both. Now since the bodies we have mentioned are in equilibrium, the cylinder with κ as its center of gravity, the sphere and the cone transferred as we have said so that they have θ as center of gravity, then $\theta \alpha : \alpha \kappa =$ cylinder : sphere + cone. But $\theta \alpha = 2 \alpha \kappa$, and hence also the cylinder $= 2 \times$ (sphere + cone). But it is also true that the cylinder $= 3$ cones [Euclid, *Elem.* XII, 10], hence 3 cones $= 2$ cones $+ 2$ spheres. If 2 cones be subtracted from both sides, then the cone whose axes form the triangle $\alpha \epsilon \zeta = 2$ spheres. But the cone whose axes form the triangle $\alpha \epsilon \zeta = 8$ cones whose axes form the triangle $\alpha \beta \delta$ because $\epsilon \zeta = 2 \beta \delta$, hence the aforesaid 8 cones $= 2$ spheres. Consequently the sphere whose greatest circle is $\alpha \beta \gamma \delta$ is four times as large as the cone with its vertex at α, and whose base is the circle on the diameter $\beta \delta$ perpendicular to $\alpha \gamma$.

Draw the straight lines $\phi\beta\chi$ and $\psi\delta\omega$ ∥ $\alpha\gamma$ through β and δ in the parallelogram $\lambda\zeta$ and imagine a cylinder whose bases are the circles on the diameters $\phi\psi$ and $\chi\omega$ and whose axis is $\alpha\gamma$. Now since the cylinder whose axes form the parallelogram $\phi\omega$ is twice as large as the cylinder whose axes form the parallelogram $\phi\delta$ and the latter is three times as large as the cone the triangle of whose axes is $\alpha\beta\delta$, as is shown in the Elements [Euclid, *Elem.* XII, 10], the cylinder whose axes form the parallelogram $\phi\omega$ is six times as large as the cone whose axes form the triangle $\alpha\beta\delta$. But it was shown that the sphere whose largest circle is $\alpha\beta\gamma\delta$ is four times as large as the same cone, consequently the cylinder is one and one half times as large as the sphere, Q. E. D.

After I had thus perceived that a sphere is four times as large as the cone whose base is the largest circle of the sphere and whose altitude is equal to its radius, it occurred to me that the surface of a sphere is four times as great as its largest circle, in which I proceeded from the idea that just as a circle is equal to a triangle whose base is the periphery of the circle and whose altitude is equal to its radius, so a sphere is equal to a cone whose base is the same as the surface of the sphere and whose altitude is equal to the radius of the sphere.

III.

By this method it may also be seen that a cylinder whose base is equal to the largest circle of a spheroid and whose altitude is equal to the axis of the spheroid, is one and one half times as large as the spheroid, and when this is recognized it becomes clear that if a spheroid is cut through its center by a plane perpendicular to its axis, one-half of the spheroid is twice as great as the cone whose base is that of the segment and its axis the same.

For let a spheroid be cut by a plane through its axis and let there be in its surface an ellipse $\alpha\beta\gamma\delta$ [Fig. 3] whose diameters are $\alpha\gamma$ and $\beta\delta$ and whose center is κ and let there be a circle in the spheroid on the diameter $\beta\delta$ perpendicular to $\alpha\gamma$; then imagine a cone whose base is the same circle but whose vertex is at α, and producing its surface, let the cone be cut by a plane through γ parallel to the base; the intersection will be a circle perpendicular to $\alpha\gamma$ with $\epsilon\zeta$ as its diameter. Now imagine a cylinder whose base is the same circle with the diameter $\epsilon\zeta$ and whose axis is $\alpha\gamma$; let $\gamma\alpha$ be produced so that $\alpha\theta = \gamma\alpha$; think of $\theta\gamma$ as a scale-beam with its center at α and in the parallelogram $\lambda\theta$ draw a straight line $\mu\nu \parallel \epsilon\zeta$, and on $\mu\nu$ construct a plane perpendicular to $\alpha\gamma$; this will intersect the cylinder in a circle whose diameter is $\mu\nu$, the spheroid in a circle whose diameter is ξo and the cone in a circle whose diameter is $\pi\rho$. Because $\gamma\alpha : \alpha\sigma = \epsilon\alpha : \alpha\pi = \mu\sigma : \sigma\pi$, and $\gamma\alpha = \alpha\theta$, therefore $\theta\alpha : \alpha\sigma = \mu\sigma : \sigma\pi$. But $\mu\sigma : \sigma\pi = \mu\sigma^2 : \mu\sigma \times \sigma\pi$ and $\mu\sigma \times \sigma\pi = \pi\sigma^2 + \sigma\xi^2$, for $\alpha\sigma \times \sigma\gamma : \sigma\xi^2 = \alpha\kappa \times \kappa\gamma : \kappa\beta^2 = \alpha\kappa^2 : \kappa\beta^2$ (for both ratios are equal to the ratio between the diameter and the parameter [Apollonius, Con. I, 21]) $= \alpha\sigma^2 : \sigma\pi^2$ therefore $\alpha\sigma^2 : \alpha\sigma \times \sigma\gamma = \pi\sigma^2 : \sigma\xi^2 = \sigma\pi^2 : \sigma\pi \times \pi\mu$, consequently $\mu\pi \times \pi\sigma = \sigma\xi^2$. If $\pi\sigma^2$ is added to both sides then

GEOMETRICAL SOLUTIONS DERIVED FROM MECHANICS. 21

$\mu\sigma \times \sigma\pi = \pi\sigma^2 + \sigma\xi^2$. Therefore $\theta\alpha : \alpha\sigma = \mu\sigma^2 : \pi\sigma^2 + \sigma\xi^2$. But $\mu\sigma^2 : \sigma\xi^2 + \sigma\pi^2 =$ the circle in the cylinder whose diameter is $\mu\nu$: the circle with the diameter $\xi o +$ the circle with the diameter $\pi\rho$; hence the circle whose diameter is $\mu\nu$ will in its present position be in equilibrium at the point α with the two circles whose diameters are ξo and $\pi\rho$ when they are transferred and so arranged on the scale-beam at the point α that θ is the center of gravity of both; and θ is the center of gravity of the two circles combined whose diameters are ξo and $\pi\rho$ when their position is changed, hence $\theta\alpha : \alpha\sigma =$ the

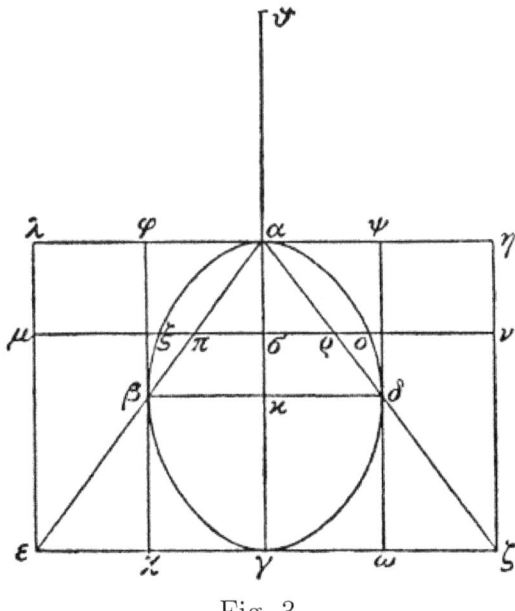

Fig. 3.

circle with the diameter $\mu\nu$: the two circles whose diameters are ξo and $\pi\rho$. In the same way it can be shown that if another straight line is drawn in the parallelogram $\lambda\zeta \parallel \epsilon\zeta$ and on this line last drawn a plane is constructed perpendicular to $\alpha\gamma$, then likewise the circle produced in the cylinder will in its present position be in equilibrium at the point α with the two circles combined which have been produced in the spheroid and in the cone respectively when they are so transferred to the point θ on the scale-beam that θ is the center of gravity of both. Then if cylinder, spheroid and cone are filled with such circles, the cylinder in its present position will be in equilibrium at the point α with the spheroid + the cone if they are transferred and so arranged on the scale-beam at the point α that θ is the center of gravity of both. Now κ is the center of gravity of the cylinder, but θ, as has been said, is the center of gravity of the spheroid and cone together. Therefore $\theta\alpha : \alpha\kappa =$ cylinder : spheroid + cone. But $\alpha\theta = 2\alpha\kappa$, hence also the cylinder $= 2 \times$ (spheroid + cone) $= 2 \times$ spheroid $+ 2 \times$ cone. But the cylinder $= 3 \times$ cone, hence $3 \times$ cone $= 2 \times$ cone $+ 2 \times$ spheroid. Subtract $2 \times$ cone from both sides; then a cone whose axes form the triangle $\alpha\epsilon\zeta = 2 \times$ spheroid. But the same cone $= 8$ cones whose axes form the $\triangle\alpha\beta\delta$; hence 8 such cones $= 2 \times$ spheroid, $4 \times$ cone $=$ spheroid; whence it follows that a spheroid is four times as great as a cone whose vertex is at α, and whose base is the circle on the diameter $\beta\delta$ perpendicular to $\lambda\epsilon$, and one-half the spheroid is twice as great as the same cone.

In the parallelogram $\lambda\zeta$ draw the straight lines $\phi\chi$ and

GEOMETRICAL SOLUTIONS DERIVED FROM MECHANICS. 23

$\psi\omega \parallel \alpha\gamma$ through the points β and δ and imagine a cylinder whose bases are the circles on the diameters $\phi\psi$ and $\chi\omega$, and whose axis is $\alpha\gamma$. Now since the cylinder whose axes form the parallelogram $\phi\omega$ is twice as great as the cylinder whose axes form the parallelogram $\phi\delta$ because their bases are equal but the axis of the first is twice as great as the axis of the second, and since the cylinder whose axes form the parallelogram $\phi\delta$ is three times as great as the cone whose vertex is at α and whose base is the circle on the diameter $\beta\delta$ perpendicular to $\alpha\gamma$, then the cylinder whose axes form the parallelogram $\phi\omega$ is six times as great as the aforesaid cone. But it has been shown that the spheroid is four times as great as the same cone, hence the cylinder is one and one half times as great as the spheroid. Q. E. D.

IV.

That a segment of a right conoid cut by a plane perpendicular to its axis is one and one half times as great as the cone having the same base and axis as the segment, can be proved by the same method in the following way:

Let a right conoid be cut through its axis by a plane intersecting the surface in a parabola $\alpha\beta\gamma$ [Fig. 4]; let it be also cut by another plane perpendicular to the axis, and let their common line of intersection be $\beta\gamma$. Let the axis of the segment be $\delta\alpha$ and let it be produced to θ so that $\theta\alpha = \alpha\delta$. Now imagine $\delta\theta$ to be a scale-beam with its center at α; let the base of the segment be the circle on the diameter $\beta\gamma$

perpendicular to $\alpha\delta$; imagine a cone whose base is the circle on the diameter $\beta\gamma$, and whose vertex is at α. Imagine also a cylinder whose base is the circle on the diameter $\beta\gamma$ and its axis $\alpha\delta$, and in the parallelogram let a straight line $\mu\nu$ be drawn $\parallel \beta\gamma$ and on $\mu\nu$ construct a plane perpendicular to $\alpha\delta$; it will intersect the cylinder in a circle whose diameter is $\mu\nu$, and the segment of the right conoid in a circle whose diameter is ξo. Now since $\beta\alpha\gamma$ is a parabola, $\alpha\delta$ its diameter and $\xi\sigma$ and $\beta\delta$ its ordinates, then [*Quadr. parab.* 3] $\delta\alpha : \alpha\sigma = \beta\delta^2 : \xi\sigma^2$. But $\delta\alpha = \alpha\theta$, therefore $\theta\alpha : \alpha\sigma = \mu\sigma^2 : \sigma\xi^2$. But $\mu\sigma^2 : \sigma\xi^2 = $ the circle in the cylinder whose diameter is $\mu\nu : $ the circle in the segment of the right conoid whose diameter is ξo, hence $\theta\alpha : \alpha\sigma = $ the circle with the diameter $\mu\nu : $ the circle with the diameter ξo; therefore the circle in the cylinder whose diameter is $\mu\nu$ is in its present position, in equilibrium at the point α with the circle whose diameter is ξo if this be transferred and so arranged on the scale-beam at θ that θ is its center of gravity. And the center of gravity of the circle whose diameter is $\mu\nu$ is at σ, that of the circle whose diameter is ξo when its position is changed, is θ, and we have the inverse proportion, $\theta\alpha : \alpha\sigma = $ the circle with the diameter $\mu\nu : $ the circle with the diameter ξo. In the same way it can be shown that if another straight line be drawn in the parallelogram $\epsilon\gamma \parallel \beta\gamma$ the circle formed in the cylinder, will in its present position be in equilibrium at the point α with that formed in the segment of the right conoid if the latter is so transferred to θ on the scale-beam that θ is its center of gravity. Therefore if the cylinder and the

GEOMETRICAL SOLUTIONS DERIVED FROM MECHANICS. 25

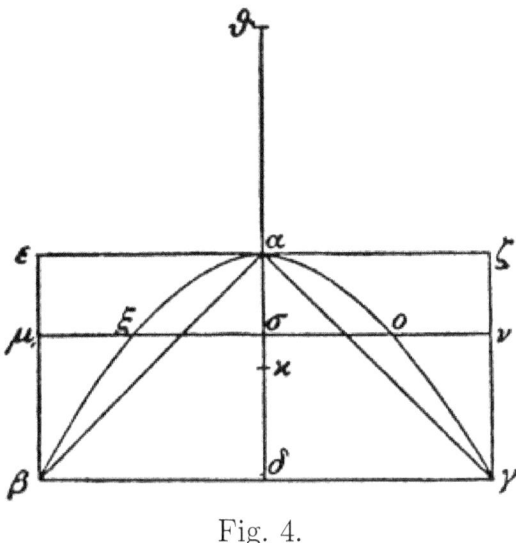

Fig. 4.

segment of the right conoid are filled up then the cylinder in its present position will be in equilibrium at the point α with the segment of the right conoid if the latter is transferred and so arranged on the scale-beam at θ that θ is its center of gravity. And since these magnitudes are in equilibrium at α, and κ is the center of gravity of the cylinder, if $\alpha\delta$ is bisected at κ and θ is the center of gravity of the segment transferred to that point, then we have the inverse proportion $\theta\alpha : \alpha\kappa =$ cylinder : segment. But $\theta\alpha = 2\alpha\kappa$ and also the cylinder $= 2 \times$ segment. But the same cylinder is 3 times as great as the cone whose base is the circle on the diameter $\beta\gamma$ and whose vertex is at α; therefore it is clear that the segment is one and one half times as great as the same cone.

V.

That the center of gravity of a segment of a right conoid which is cut off by a plane perpendicular to the axis, lies on the straight line which is the axis of the segment divided in such a way that the portion at the vertex is twice as great as the remainder, may be perceived by our method in the following way:

Let a segment of a right conoid cut off by a plane perpendicular to the axis be cut by another plane through the axis, and let the intersection in its surface be the parabola $\alpha\beta\gamma$ [Fig. 5] and let the common line of intersection of the plane which cut off the segment and of the intersecting plane be $\beta\gamma$; let the axis of the segment and the diameter of the parabola $\alpha\beta\gamma$ be $\alpha\delta$; produce $\delta\alpha$ so that $\alpha\theta = \alpha\delta$ and imagine $\delta\theta$ to be a scale-beam with its center at α; then inscribe a cone in the segment with the lateral boundaries $\beta\alpha$ and $\alpha\gamma$ and in the parabola draw a straight line $\xi o \parallel \beta\gamma$ and let it cut the parabola in ξ and o and the lateral boundaries of the cone in π and ρ. Now because $\xi\sigma$ and $\beta\delta$ are drawn perpendicular to the diameter of the parabola, $\delta\alpha : \alpha\sigma = \beta\delta^2 : \xi\sigma^2$ [*Quadr. parab.* 3]. But $\delta\alpha : \alpha\sigma = \beta\delta : \pi\sigma = \beta\delta^2 : \beta\delta \times \pi\sigma$, therefore also $\beta\delta^2 : \xi\sigma^2 = \beta\delta^2 : \beta\delta \times \pi\sigma$. Consequently $\xi\sigma^2 = \beta\delta \times \pi\sigma$ and $\beta\delta : \xi\sigma = \xi\sigma : \pi\sigma$, therefore $\beta\delta : \pi\sigma = \xi\sigma^2 : \sigma\pi^2$. But $\beta\delta : \pi\sigma = \delta\alpha : \alpha\sigma = \theta\alpha : \alpha\sigma$, therefore also $\theta\alpha : \alpha\sigma = \xi\sigma^2 : \sigma\pi^2$. On ξo construct a plane perpen-

GEOMETRICAL SOLUTIONS DERIVED FROM MECHANICS. 27

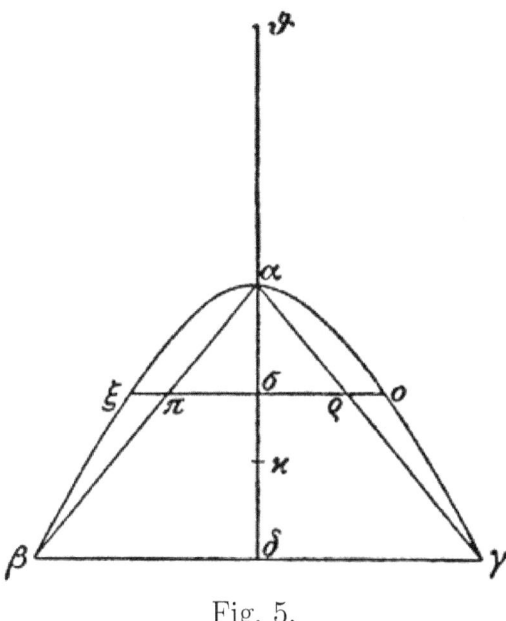

Fig. 5.

dicular to $\alpha\delta$; this will intersect the segment of the right conoid in a circle whose diameter is ξo and the cone in a circle whose diameter is $\pi\rho$. Now because $\theta\alpha : \alpha\sigma = \xi\sigma^2 : \sigma\pi^2$ and $\xi\sigma^2 : \sigma\pi^2 = $ the circle with the diameter ξo : the circle with the diameter $\pi\rho$, therefore $\theta\alpha : \alpha\sigma = $ the circle whose diameter is ξo : the circle whose diameter is $\pi\rho$. Therefore the circle whose diameter is ξo will in its present position be in equilibrium at the point α with the circle whose diameter is $\pi\rho$ when this is so transferred to θ on the scale-beam that θ is its center of gravity. Now since σ is the center of gravity

GEOMETRICAL SOLUTIONS DERIVED FROM MECHANICS.   28

of the circle whose diameter is ξo in its present position, and θ is the center of gravity of the circle whose diameter is $\pi\rho$ if its position is changed as we have said, and inversely $\theta\alpha : \alpha\sigma =$ the circle with the diameter ξo : the circle with the diameter $\pi\rho$, then the circles are in equilibrium at the point α. In the same way it can be shown that if another straight line is drawn in the parabola $\parallel \beta\gamma$ and on this line last drawn a plane is constructed perpendicular to $\alpha\delta$, the circle formed in the segment of the right conoid will in its present position be in equilibrium at the point α with the circle formed in the cone, if the latter is transferred and so arranged on the scale-beam at θ that θ is its center of gravity. Therefore if the segment and the cone are filled up with circles, all circles in the segment will be in their present positions in equilibrium at the point α with all circles of the cone if the latter are transferred and so arranged on the scale-beam at the point θ that θ is their center of gravity. Therefore also the segment of the right conoid in its present position will be in equilibrium at the point α with the cone if it is transferred and so arranged on the scale-beam at θ that θ is its center of gravity. Now because the center of gravity of both magnitudes taken together is α, but that of the cone alone when its position is changed is θ, then the center of gravity of the remaining magnitude lies on $\alpha\theta$ extended towards α if $\alpha\kappa$ is cut off in such a way that $\alpha\theta : \alpha\kappa =$ segment : cone. But the segment is one and one half the size of the cone, consequently $\alpha\theta = \frac{3}{2}\alpha\kappa$ and κ, the center of gravity of the right conoid, so divides $\alpha\delta$ that the

GEOMETRICAL SOLUTIONS DERIVED FROM MECHANICS.   29

portion at the vertex of the segment is twice as large as the remainder.

VI.

[The center of gravity of a hemisphere is so divided on its axis] that the portion near the surface of the hemisphere is in the ratio of 5 : 3 to the remaining portion.

Let a sphere be cut by a plane through its center intersecting the surface in the circle $\alpha\beta\gamma\delta$ [Fig. 6], $\alpha\gamma$ and $\beta\delta$ being two diameters of the circle perpendicular to each other. Let a plane be constructed on $\beta\delta$ perpendicular to $\alpha\gamma$. Then imagine a cone whose base is the circle with the diameter $\beta\delta$, whose vertex is at α and its lateral boundaries are $\beta\alpha$ and $\alpha\delta$; let $\gamma\alpha$ be produced so that $\alpha\theta = \gamma\alpha$, imagine the straight line $\theta\gamma$ to be a scale-beam with its center at α and in the semicircle $\beta\alpha\delta$ draw a straight line $\xi o \parallel \beta\delta$; let it cut the circumference of the semicircle in ξ and o, the lateral boundaries of the cone in π and ρ, and $\alpha\gamma$ in ϵ. On ξo construct a plane perpendicular to $\alpha\epsilon$; it will intersect the hemisphere in a circle with the diameter ξo, and the cone in a circle with the diameter $\pi\rho$. Now because $\alpha\gamma : \alpha\epsilon = \xi\alpha^2 : \alpha\epsilon^2$ and $\xi\alpha^2 = \alpha\epsilon^2 + \epsilon\xi^2$ and $\alpha\epsilon = \epsilon\pi$, therefore $\alpha\gamma : \alpha\epsilon = \xi\epsilon^2 + \epsilon\pi^2 : \epsilon\pi^2$. But $\xi\epsilon^2 + \epsilon\pi^2 : \epsilon\pi^2 =$ the circle with the diameter $\xi o +$ the circle with the diameter $\pi\rho$: the circle with the diameter $\pi\rho$, and $\gamma\alpha = \alpha\theta$, hence $\theta\alpha : \alpha\epsilon =$ the circle with the diameter $\xi o +$ the circle with the diameter $\pi\rho$: circle with the diameter $\pi\rho$. Therefore the

GEOMETRICAL SOLUTIONS DERIVED FROM MECHANICS. 30

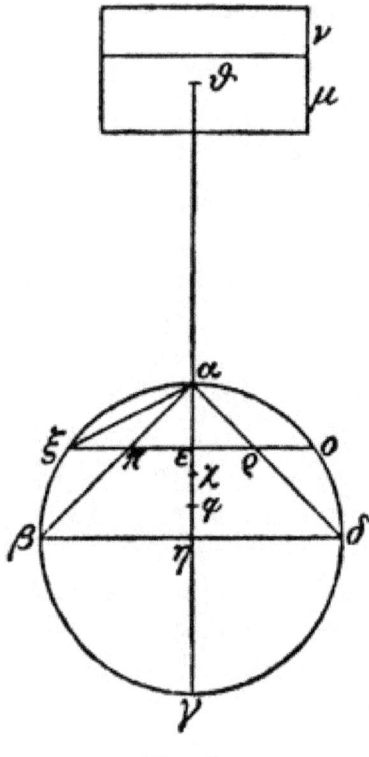

Fig. 6.

two circles whose diameters are ξo and πρ in their present position are in equilibrium at the point α with the circle whose diameter is πρ if it is transferred and so arranged at θ that θ is its center of gravity. Now since the center of gravity of the two circles whose diameters are ξo and πρ in their present position [is the point ϵ, but of the circle whose diameter is $\pi\rho$ when its position is changed is the point θ, then $\theta\alpha : \alpha\epsilon =$ the circles whose diameters are] ξo[, $\pi\rho$: the circle whose diameter is $\pi\rho$. In the same way if another straight line in the] hemisphere $\beta\alpha\delta$ [is drawn $\parallel \beta\delta$ and a plane is constructed] perpendicular to [$\alpha\gamma$ the] two [circles produced in the cone and in the hemisphere are in their position] in equilibrium at α [with the circle which is produced in the cone] if it is transferred and arranged on the scale at θ. [Now if] the hemisphere and the cone [are filled up with circles then all circles in the] hemisphere and those [in the cone] will in their present position be in equilibrium [with all circles] in the cone, if these are transferred and so arranged on the scale-beam at θ that θ is their center of gravity; [therefore the hemisphere and cone also] are in their position [in equilibrium at the point α] with the cone if it is transferred and so arranged [on the scale-beam at θ] that θ is its center of gravity.

VII.

By [this method] it may also be perceived that [any segment whatever] of a sphere bears the same ratio to a cone having the same [base] and axis [that the radius of the sphere + the axis of the opposite segment : the axis of the opposite segment] ....
and [Fig. 7] on $\mu\nu$ construct a plane perpendicular to $\alpha\gamma$; it will intersect the cylinder in a circle whose diameter is $\mu\nu$,

the segment of the sphere in a circle whose diameter is ξο

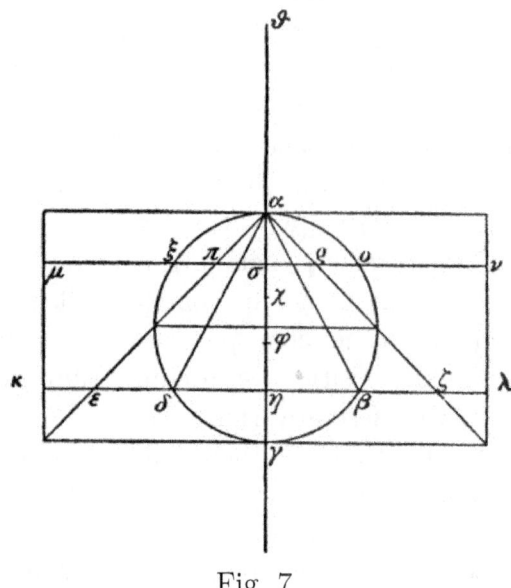

Fig. 7.

and the cone whose base is the circle on the diameter εζ and whose vertex is at α in a circle whose diameter is πρ. In the same way as before it may be shown that a circle whose diameter is μν is in its present position in equilibrium at α with the two circles [whose diameters are ξο and πρ if they are so arranged on the scale-beam that θ is their center of gravity. [And the same can be proved of all corresponding circles.] Now since cylinder, cone, and spherical segment are filled up with such circles, the cylinder in its present position [will be in equilibrium at α] with the cone + the

GEOMETRICAL SOLUTIONS DERIVED FROM MECHANICS. 33

spherical segment if they are transferred and attached to the scale-beam at θ. Divide $\alpha\eta$ at ϕ and χ so that $\alpha\chi = \chi\eta$ and $\eta\phi = \frac{1}{3}\alpha\phi$; then χ will be the center of gravity of the cylinder because it is the center of the axis $\alpha\eta$. Now because the above mentioned bodies are in equilibrium at α, cylinder : cone with the diameter of its base $\epsilon\zeta$ + the spherical segment $\beta\alpha\delta = \theta\alpha : \alpha\chi$. And because $\eta\alpha = 3\eta\phi$ then $[\gamma\eta \times \eta\phi] = \frac{1}{3}\alpha\eta \times \eta\gamma$. Therefore also $\gamma\eta \times \eta\phi = \frac{1}{3}\beta\eta^2$. ....

VIIa.

In the same way it may be perceived that any segment of an ellipsoid cut off by a perpendicular plane, bears the same ratio to a cone having the same base and the same axis, as half of the axis of the ellipsoid + the axis of the opposite segment bears to the axis of the opposite segment. ....

VIII.

....

produce $\alpha\gamma$ [Fig. 8] making $\alpha\theta = \alpha\gamma$ and $\gamma\xi$ = the radius of the sphere; imagine $\gamma\theta$ to be a scale-beam with a center at α, and in the plane cutting off the segment inscribe a circle with its center at η and its radius = $\alpha\eta$; on this circle construct a cone with its vertex at α and its lateral boundaries $\alpha\epsilon$ and $\alpha\zeta$. Then draw a straight line $\kappa\lambda \parallel \epsilon\zeta$; let it cut the circumference of the segment at κ and λ, the lateral boundaries of the cone $\alpha\epsilon\zeta$ at ρ and o and $\alpha\gamma$ at π. Now because

GEOMETRICAL SOLUTIONS DERIVED FROM MECHANICS.   34

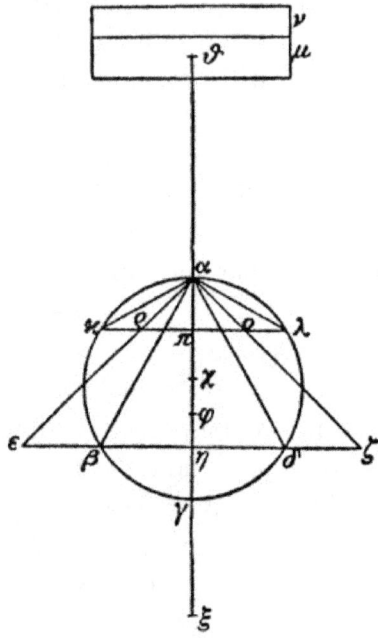

Fig. 8.

$\alpha\gamma : \alpha\pi = \alpha\kappa^2 : \alpha\pi^2$ and $\kappa\alpha^2 = \alpha\pi^2 + \pi\kappa^2$ and $\alpha\pi^2 = \pi o^2$ (since also $\alpha\eta^2 = \epsilon\eta^2$), then $\gamma\alpha : \alpha\pi = \kappa\pi^2 + \pi o^2 : o\pi^2$. But $\kappa\pi^2 + \pi o^2 : \pi o^2 =$ the circle with the diameter $\kappa\lambda +$ the circle with the diameter $o\rho$: the circle with the diameter $o\rho$ and $\gamma\alpha = \alpha\theta$; therefore $\theta\alpha : \alpha\pi =$ the circle with the diameter $\kappa\lambda +$ the circle with the diameter $o\rho$: the circle with the diameter $o\rho$. Now since the circle with the diameter $\kappa\lambda +$ the circle with the diameter $o\rho$: the circle with the diam-

GEOMETRICAL SOLUTIONS DERIVED FROM MECHANICS. 35

eter $o\rho = \alpha\theta : \pi\alpha$, let the circle with the diameter $o\rho$ be transferred and so arranged on the scale-beam at θ that θ is its center of gravity; then $\theta\alpha : \alpha\pi =$ the circle with the diameter $\kappa\lambda +$ the circle with the diameter $o\rho$ in their present positions : the circle with the diameter $o\rho$ if it is transferred and so arranged on the scale-beam at θ that θ is its center of gravity. Therefore the circles in the segment $\beta\alpha\delta$ and in the cone $\alpha\epsilon\zeta$ are in equilibrium at α with that in the cone $\alpha\epsilon\zeta$. And in the same way all circles in the segment $\beta\alpha\delta$ and in the cone $\alpha\epsilon\zeta$ in their present positions are in equilibrium at the point α with all circles in the cone $\alpha\epsilon\zeta$ if they are transferred and so arranged on the scale-beam at θ that θ is their center of gravity; then also the spherical segment $\alpha\beta\delta$ and the cone $\alpha\epsilon\zeta$ in their present positions are in equilibrium at the point α with the cone $\epsilon\alpha\zeta$ if it is transferred and so arranged on the scale-beam at θ that θ is its center of gravity. Let the cylinder $\mu\nu$ equal the cone whose base is the circle with the diameter $\epsilon\zeta$ and whose vertex is at α and let $\alpha\eta$ be so divided at ϕ that $\alpha\eta = 4\phi\eta$; then ϕ is the center of gravity of the cone $\epsilon\alpha\zeta$ as has been previously proved. Moreover let the cylinder $\mu\nu$ be so cut by a perpendicularly intersecting plane that the cylinder μ is in equilibrium with the cone $\epsilon\alpha\zeta$. Now since the segment $\alpha\beta\delta +$ the cone $\epsilon\alpha\zeta$ in their present positions are in equilibrium at α with the cone $\epsilon\alpha\zeta$ if it is transferred and so arranged on the scale-beam at θ that θ is its center of gravity, and cylinder $\mu\nu =$ cone $\epsilon\alpha\zeta$ and the two cylinders $\mu + \nu$ are moved to θ and $\mu\nu$ is in equilibrium with both bodies, then will also the cylinder ν be in equilibrium with the segment of the sphere at the point α. And since the spherical segment $\beta\alpha\delta$: the cone whose base is the circle with the diameter $\beta\delta$, and whose vertex is at $\alpha = \xi\eta : \eta\gamma$ (for this has previously been proved [*De sph. et cyl.* II, 2 Coroll.]) and cone $\beta\alpha\delta$: cone $\epsilon\alpha\zeta$ = the circle with the diameter $\beta\delta$: the circle with the diameter $\epsilon\zeta = \beta\eta^2 : \eta\epsilon^2$, and $\beta\eta^2 = \gamma\eta \times \eta\alpha$, $\eta\epsilon^2 = \eta\alpha^2$, and $\gamma\eta \times \eta\alpha : \eta\alpha^2 = \gamma\eta : \eta\alpha$, therefore cone $\beta\alpha\delta$: cone $\epsilon\alpha\zeta = \gamma\eta : \eta\alpha$. But we have shown that cone $\beta\alpha\delta$: segment $\beta\alpha\delta = \gamma\eta : \eta\xi$, hence δι' ἴσου segment $\beta\alpha\delta$: cone $\epsilon\alpha\zeta = \xi\eta : \eta\alpha$. And because $\alpha\chi : \chi\eta = \eta\alpha+4\eta\gamma : \alpha\eta+2\eta\gamma$ so inversely $\eta\chi : \chi\alpha = 2\gamma\eta+\eta\alpha : 4\gamma\eta+\eta\alpha$ and by addition $\eta\alpha : \alpha\chi = 6\gamma\eta + 2\eta\alpha : \eta\alpha + 4\eta\gamma$. But $\eta\xi = \frac{1}{4}(6\eta\gamma + 2\eta\alpha)$ and $\gamma\phi = \frac{1}{4}(4\eta\gamma + \eta\alpha)$; for that is evident. Hence $\eta\alpha : \alpha\chi = \xi\eta : \gamma\phi$, consequently also $\xi\eta : \eta\alpha = \gamma\phi : \chi\alpha$. But it was also demonstrated that $\xi\eta : \eta\alpha$ = the segment whose vertex is at α and whose base is the circle with the diameter $\beta\delta$: the cone whose vertex is at α and whose base is the circle with the diameter $\epsilon\zeta$; hence segment $\beta\alpha\delta$: cone $\epsilon\alpha\zeta = \gamma\phi : \chi\alpha$. And since the cylinder μ is in equilibrium with the cone $\epsilon\alpha\zeta$ at α, and θ is the center of gravity of the cylinder while ϕ is that of the cone $\epsilon\alpha\zeta$, then cone $\epsilon\alpha\zeta$: cylinder $\mu = \theta\alpha : \alpha\phi = \gamma\alpha : \alpha\phi$. But cylinder $\mu\nu$ = cone $\epsilon\alpha\zeta$; hence by subtraction, cylinder μ : cylinder $\nu = \alpha\phi : \gamma\phi$. And cylinder $\mu\nu$ = cone $\epsilon\alpha\zeta$; hence cone $\epsilon\alpha\zeta$: cylinder $\nu = \gamma\alpha : \gamma\phi = \theta\alpha : \gamma\phi$. But it was also demonstrated that segment $\beta\alpha\delta$: cone $\epsilon\alpha\zeta = \gamma\phi : \chi\alpha$; hence δι' ἴσου segment $\beta\alpha\delta$: cylinder $\nu = \zeta\alpha : \alpha\chi$. And it was demonstrated that segment $\beta\alpha\delta$ is in equilibrium at α

GEOMETRICAL SOLUTIONS DERIVED FROM MECHANICS. 37

with the cylinder ν and θ is the center of gravity of the cylinder ν, consequently the point χ is also the center of gravity of the segment $\beta\alpha\delta$.

IX.

In a similar way it can also be perceived that the center of gravity of any segment of an ellipsoid lies on the straight line which is the axis of the segment so divided that the portion at the vertex of the segment bears the same ratio to the remaining portion as the axis of the segment + 4 times the axis of the opposite segment bears to the axis of the segment + twice the axis of the opposite segment.

X.

It can also be seen by this method that [a segment of a hyperboloid] bears the same ratio to a cone having the same base and axis as the segment, that the axis of the segment + 3 times the addition to the axis bears to the axis of the segment of the hyperboloid + twice its addition [*De Conoid.* 25]; and that the center of gravity of the hyperboloid so divides the axis that the part at the vertex bears the same ratio to the rest that three times the axis + eight times the addition to the axis bears to the axis of the hyperboloid + 4 times the addition to the axis, and many other points which I will leave aside since the method has been made clear by the examples already given and only the

GEOMETRICAL SOLUTIONS DERIVED FROM MECHANICS.   38

demonstrations of the above given theorems remain to be stated.

XI.

When in a perpendicular prism with square bases a cylinder is inscribed whose bases lie in opposite squares and whose curved surface touches the four other parallelograms, and when a plane is passed through the center of the circle which is the base of the cylinder and one side of the opposite square, then the body which is cut off by this plane [from the cylinder] will be $\frac{1}{6}$ of the entire prism. This can be perceived through the present method and when it is so warranted we will pass over to the geometrical proof of it.

Imagine a perpendicular prism with square bases and a cylinder inscribed in the prism in the way we have described. Let the prism be cut through the axis by a plane perpendicular to the plane which cuts off the section of the cylinder; this will intersect the prism containing the cylinder in the parallelogram $\alpha\beta$ [Fig. 9] and the common intersecting line of the plane which cuts off the section of the cylinder and the plane lying through the axis perpendicular to the one cutting off the section of the cylinder will be $\beta\gamma$; let the axis of the cylinder and the prism be $\gamma\delta$ which is bisected at right angles by $\epsilon\zeta$ and on $\epsilon\zeta$ let a plane be constructed perpendicular to $\gamma\delta$. This will intersect the prism in a square and the cylinder in a circle.

Now let the intersection of the prism be the square $\mu\nu$

GEOMETRICAL SOLUTIONS DERIVED FROM MECHANICS. 39

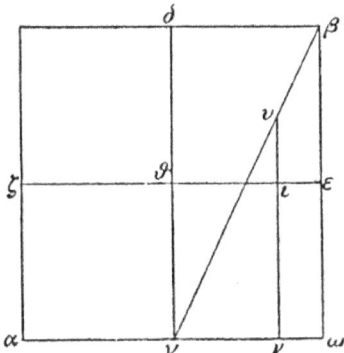 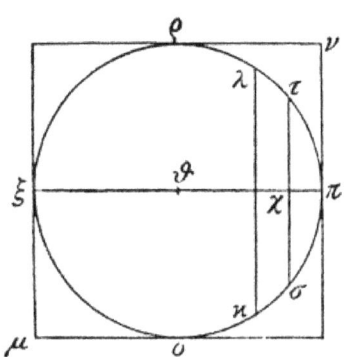

[Fig. 10], that of the cylinder, the circle ξοπρ and let the circle touch the sides of the square at the points ξ, ο, π and ρ; let the common line of intersection of the plane cutting off the cylinder-section and that passing through εζ perpendicular to the axis of the cylinder, be κλ; this line is bisected by πθξ. In the semicircle οπρ draw a straight line στ perpendicular to πχ, on στ construct a plane perpendicular to ξπ and produce it to both sides of the plane enclosing the circle ξοπρ; this will intersect the half-cylinder whose base is the semicircle οπρ and whose altitude is the axis of the prism, in a parallelogram one side of which = στ and the other = the vertical boundary of the cylinder, and it will intersect the cylinder-section likewise in a parallelogram of which one side is στ and the other μν [Fig. 9]; and accordingly μν will be drawn in the parallelogram δε || βω and will cut off ει = πχ. Now because εγ is a parallelogram and νι || θγ, and εθ and βγ cut the parallels, therefore εθ : θι = ωγ : γν = βω : νν.

But $\beta\omega : \upsilon\nu$ = parallelogram in the half-cylinder : parallelogram in the cylinder-section, therefore both parallelograms have the same side $\sigma\tau$; and $\epsilon\theta = \theta\pi$, $\iota\theta = \chi\theta$; and since $\pi\theta = \theta\xi$ therefore $\theta\xi : \theta\chi$ = parallelogram in half-cylinder : parallelogram in the cylinder-section. Imagine the parallelogram in the cylinder-section transferred and so brought to ξ that ξ is its center of gravity, and further imagine $\pi\xi$ to be a scale-beam with its center at θ; then the parallelogram in the half-cylinder in its present position is in equilibrium at the point θ with the parallelogram in the cylinder-section when it is transferred and so arranged on the scale-beam at ξ that ξ is its center of gravity. And since χ is the center of gravity in the parallelogram in the half-cylinder, and ξ that of the parallelogram in the cylinder-section when its position is changed, and $\xi\theta : \theta\chi$ = the parallelogram whose center of gravity is χ : the parallelogram whose center of gravity is ξ, then the parallelogram whose center of gravity is χ will be in equilibrium at θ with the parallelogram whose center of gravity is ξ. In this way it can be proved that if another straight line is drawn in the semicircle $\sigma\pi\rho$ perpendicular to $\pi\theta$ and on this straight line a plane is constructed perpendicular to $\pi\theta$ and is produced towards both sides of the plane in which the circle $\xi o\pi\rho$ lies, then the parallelogram formed in the half-cylinder in its present position will be in equilibrium at the point θ with the parallelogram formed in the cylinder-section if this is transferred and so arranged on the scale-beam at ξ that ξ is its center of gravity; therefore also all parallelograms in the half-cylinder in their present

GEOMETRICAL SOLUTIONS DERIVED FROM MECHANICS. 41

positions will be in equilibrium at the point θ with all parallelograms of the cylinder-section if they are transferred and attached to the scale-beam at the point ξ; consequently also the half-cylinder in its present position will be in equilibrium at the point θ with the cylinder-section if it is transferred and so arranged on the scale-beam at ξ that ξ is its center of gravity.

XII.

Let the parallelogram $\mu\nu$ be perpendicular to the axis [of the circle] ξo [$\pi\rho$] [Fig. 11]. Draw $\theta\mu$ and $\theta\eta$ and erect upon them two planes perpendicular to the plane in which the semicircle $o\pi\rho$ lies and extend these planes on both sides. The result is a prism whose base is a triangle similar to $\theta\mu\eta$ and whose altitude is equal to the axis of the cylinder, and this prism is $\frac{1}{4}$ of the entire prism which contains the cylinder. In the semicircle $o\pi\rho$ and in the square $\mu\nu$ draw two straight lines $\kappa\lambda$ and $\tau\upsilon$ at equal distances from $\pi\xi$; these will cut the circumference of the semicircle $o\pi\rho$ at the points κ and τ, the diameter $o\rho$ at σ and ζ and the straight lines $\theta\eta$ and $\theta\mu$ at ϕ and χ. Upon $\kappa\lambda$ and $\tau\upsilon$ construct two planes perpendicular to $o\rho$ and extend them towards both sides of the plane in which lies the circle $\xi o\pi\rho$; they will intersect the half-cylinder whose base is the semicircle $o\pi\rho$ and whose altitude is that of the cylinder, in a parallelogram one side of which $= \kappa\sigma$ and the other $=$ the axis of the cylinder; and they will intersect the prism $\theta\eta\mu$ likewise in a parallelogram

GEOMETRICAL SOLUTIONS DERIVED FROM MECHANICS. 42

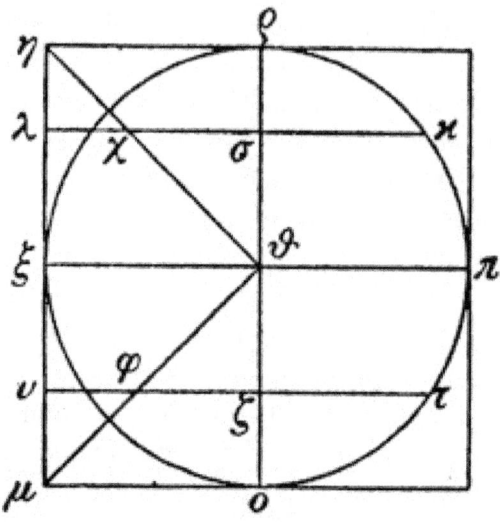

Fig. 11.

one side of which is equal to $\lambda\chi$ and the other equal to the axis, and in the same way the half-cylinder in a parallelogram one side of which $= \tau\zeta$ and the other $=$ the axis of the cylinder, and the prism in a parallelogram one side of which $= \nu\phi$ and the other $=$ the axis of the cylinder. . . . .

XIII.

Let the square $\alpha\beta\gamma\delta$ [Fig. 12] be the base of a perpendicular prism with square bases and let a cylinder be inscribed in the prism whose base is the circle $\epsilon\zeta\eta\theta$ which touches the sides of the parallelogram $\alpha\beta\gamma\delta$ at ϵ, ζ, η, and θ. Pass a

plane through its center and the side in the square opposite the square $\alpha\beta\gamma\delta$ corresponding to the side $\gamma\delta$; this will cut off from the whole prism a second prism which is $\frac{1}{4}$ the size of the whole prism and which will be bounded by three parallelograms and two opposite triangles. In the semicircle $\epsilon\zeta\eta$ describe a parabola whose origin is $\eta\epsilon$ and whose axis is $\zeta\kappa$, and in the parallelogram $\delta\eta$ draw $\mu\nu \parallel \kappa\zeta$; this will cut the circumference of the semicircle at ξ, the parabola at λ, and $\mu\nu \times \nu\lambda = \nu\zeta^2$ (for this is evident [Apollonios, Con. I, 11]). Therefore $\mu\nu : \nu\lambda = \kappa\eta^2 : \lambda\sigma^2$. Upon $\mu\nu$ construct a plane parallel to $\epsilon\eta$; this will intersect the prism cut off from the whole prism in a right-angled triangle one side of which is $\mu\nu$ and the other a straight line in the plane upon $\gamma\delta$ perpendicular to $\gamma\delta$ at ν and equal to the axis of the cylinder, but whose hypotenuse is in the intersecting plane. It will intersect the portion which is cut off from the cylinder by the plane passed through $\epsilon\eta$ and the side of the square opposite the side $\gamma\delta$ in a right-angled triangle one side of which is $\mu\xi$ and the other a straight line drawn in the surface of the cylinder perpendicular to the plane $\kappa\nu$, and the hypotenuse ....
and all the triangles in the prism : all the triangles in the cylinder-section = all the straight lines in the parallelogram $\delta\eta$: all the straight lines between the parabola and the straight line $\epsilon\eta$. And the prism consists of the triangles in the prism, the cylinder-section of those in the cylinder-section, the parallelogram $\delta\eta$ of the straight lines in the parallelogram $\delta\eta \parallel \kappa\zeta$ and the segment of the parabola of the straight

GEOMETRICAL SOLUTIONS DERIVED FROM MECHANICS. 44

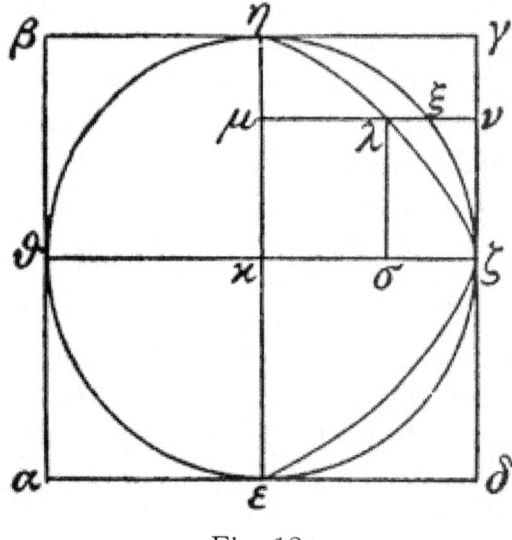

Fig. 12.

lines cut off by the parabola and the straight line $\epsilon\eta$; hence prism : cylinder-section = parallelogram $\eta\delta$: segment $\epsilon\zeta\eta$ that is bounded by the parabola and the straight line $\epsilon\eta$. But the parallelogram $\delta\eta = \frac{3}{2}$ the segment bounded by the parabola and the straight line $\epsilon\eta$ as indeed has been shown in the previously published work, hence also the prism is equal to one and one half times the cylinder-section. Therefore when the cylinder-section = 2, the prism = 3 and the whole prism containing the cylinder equals 12, because it is four times the size of the other prism; hence the cylinder-section is equal to $\frac{1}{6}$ of the prism, Q. E. D.

XIV.

[Inscribe a cylinder in] a perpendicular prism with square bases [and let it be cut by a plane passed through the center of the base of the cylinder and one side of the opposite square.] Then this plane will cut off a prism from the whole prism and a portion of the cylinder from the cylinder. It may be proved that the portion cut off from the cylinder by the plane is one-sixth of the whole prism. But first we will prove that it is possible to inscribe a solid figure in the cylinder-section and to circumscribe another composed of prisms of equal altitude and with similar triangles as bases, so that the circumscribed figure exceeds the inscribed less than any given magnitude. ....

But it has been shown that the prism cut off by the inclined plane $< \frac{3}{2}$ the body inscribed in the cylinder-section. Now the prism cut off by the inclined plane : the body inscribed in the cylinder-section = parallelogram $\delta\eta$: the parallelograms which are inscribed in the segment bounded by the parabola and the straight line $\epsilon\eta$. Hence the parallelogram $\delta\eta < \frac{3}{2}$ the parallelograms in the segment bounded by the parabola and the straight line $\epsilon\eta$. But this is impossible because we have shown elsewhere that the parallelogram $\delta\eta$ is one and one half times the segment bounded by the parabola and the straight line $\epsilon\eta$, consequently is ....
not greater ....

And all prisms in the prism cut off by the inclined plane : all prisms in the figure described around the cylinder-section = all parallelograms in the parallelogram $\delta\eta$: all parallelograms in the figure which is described around the segment bounded by the parabola and the straight line $\epsilon\eta$, i. e., the prism cut off by the inclined plane : the figure described around the cylinder-section = parallelogram $\delta\eta$: the figure bounded by the parabola and the straight line $\epsilon\eta$. But the prism cut off by the inclined plane is greater than one and one half times the solid figure circumscribed around the cylinder-section ....

www.ingramcontent.com/pod-product-compliance
Lightning Source LLC
Chambersburg PA
CBHW061227180526
4517OCB00003B/1198